LIFE SCIENCE IN DEPTH

ADAPTATION AND COMPETITION

Ann Fullick

www.heinemann.co.uk/library
Visit our website to find out more information about Heinemann Library books.

To order:

- Phone 44 (0) 1865 888066
- Send a fax to 44 (0) 1865 314091
- Visit the Heinemann bookshop at www.heinemann.co.uk/library to browse our catalogue and order online.

First published in Great Britain by Heinemann Library, Halley Court, Jordan Hill, Oxford OX2 8EJ, part of Harcourt Education.

Editorial: Sarah Shannon and Dave Harris
Design: Richard Parker and Q2A Solutions
Illustrations: Q2A Solutions
Picture Research: Natalie Gray
Production: Chloe Bloom

Originated by Modern Age Repro
Printed and bound in China by South China Printing Company

10 digit ISBN: 0 431 10901 X
13 digit ISBN: 978 0 431 10901 5

10 09 08 07 06
10 9 8 7 6 5 4 3 2 1

British Library Cataloguing in Publication Data
Fullick, Ann, 1956-
Adaptation and competition.
- (Life science in depth)
578.4
A full catalogue record for this book is available from the British Library.

Acknowledgements
The publishers would like to thank the following for permission to reproduce photographs:
Alamy p. **19**; Ant Photo Library p. **56** (Jack Cameron); Corbis pp. **32** (Andrew Holbrooke), **55** (Dale C. Spartas), **27** (David Muench), **49** (Gary Braasch), **24** (John Conrad), **28** (Macduff Everton), **1**, **34** (Pat O'Hara), **38** (Phil Schermeister), **4** (Photo Library Int.), **54** (Stephen Frink), **39** (Steven Doi/Zuma), **11** (Wolfgang Kaehler), **9**; Dr Peter Scott pp. **17a**, **17b**; Getty Images pp. **5** (Botanica), **53** (Digital Vision), **58** (PhotoDisc), **40** (Roine Magnusson), **7** (Thomas Schmitt); Harcourt Education Ltd pp. **12** (BrandX), **8**, **42** (Digital Vision); James Cook p. **23**; NHPA pp. **41** (Daniel Heuclin), **51** (Hellio & Van Ingen), **46** (T Kitchin & V Hurst); Oxford Scientific p. **28** (Mark Hamblin); Science Photo Library pp. **30** (Gregory Dimijian), **21** (Ray Coleman), **20** (Rod Plank); Still Pictures p. **35** (Fred Bruemmer).

Cover photograph of a chameleon, reproduced with permission of Getty Images.

Our thanks to Emma Leatherbarrow for her assistance in the preparation of this book.

The paper used to print this book comes from sustainable resources.

Contents

Words printed in the text in bold, **like this**, are explained in the Glossary.

A world of difference

The planet Earth seen from space is beautiful. Get closer, and you begin to see just how amazing it really is. There are deep oceans, high mountains, flat plains, rivers, and hills – an almost endless variety of places to live. The conditions vary too, from freezing to gentle warm breezes, through burning heat, and torrential rains to almost no rain at all. What is more, all over the surface of the Earth, making the best of whatever conditions are there, you will find living **organisms**. Plants, animals, **fungi**, **bacteria**, and **protists** – they all survive and thrive from the icy Arctic regions to the tropical forests, from the prairies to the Alps. To survive successfully in so many different places, living organisms have special features that make them suited to the conditions where they live. These special features are known as **adaptations**, and they develop over the years through the process of **natural selection**.

The Earth – home to us and millions of other living organisms.

RELATIONSHIPS OF LIFE

All living things depend on each other, and these links soon become clear. Almost all life on Earth depends on plants because plants can make food in a process known as **photosynthesis**, using the energy from sunlight to combine carbon dioxide and water to make sugars. Animals cannot photosynthesise, so they feed on plants or other animals, while the bacteria and fungi consume the bodies of both plants and animals – living and dead. As the bodies of animals and plants decompose, minerals are returned to the soil, which in turn are taken up and used by plants.

It all sounds very simple, but animals and plants often have to compete with each other to get enough to eat. Some living organisms have developed in very strange ways to give themselves a winning edge in these competitions. There are plants that live hundreds of feet up the branches of trees with roots that dangle into thin air, mammals that live their whole lives below ground and no longer have any eyes, and insects that go through their whole lifecycle inside a single flower. The natural world is full of unbelievable organisms, and in this book you are going to find out more about some of them.

The many layers of a tropical rainforest provide a home for a wide variety of living organisms, each adapted to live and feed in a particular way.

Did you know..?

No one is quite sure exactly how many different **species** of living things there are on the Earth today. Estimates range from 5 million to 30 million species – whatever the actual number is, it is a lot of life!

Your habitat, your home

An **ecosystem** is a place that can support life. The whole world is one enormous ecosystem (sometimes called the **biosphere**), but we usually look at a small part at a time, finding out about the animals and plants that live there. This smaller ecosystem might be a desert, a seashore, a meadow, even your back garden. But to make sense of what we find, we need to think about the **environment** as a whole and then the **habitat** of each living thing. The habitat of an organism is the place where it lives – its home.

ENVIRONMENTS AND HABITATS

The environment describes what a place is like. Is it rocky? What is the soil like? How strong are the tides? These are the sorts of questions that need answering. Describing the environment includes the weather conditions, the temperature, how much water there is, and what the oxygen and carbon dioxide levels are like. This sort of information is really useful – if we know what the environment is like, we have some idea of the type of living things we might find there.

But ecosystems and environments are still on quite a big scale. Think of a city as an ecosystem. Every one who lives there shares the same environment in terms of weather, roads, water supply, and so on. But living in the city is not the same for everyone. If you live in a big house in the suburbs, your life will be very different to someone living in a big estate of houses, or in a high-rise apartment. So to help us really understand the animals and plants we see around us, we need to look very closely at where they live. Within any ecosystem, there will be many different habitats, all inhabited by very different living organisms.

HOMES IN YOUR BACK GARDEN

A garden is a familiar setting, but how many of us stop to think about all the different habitats and living things it contains? A lawn is home to grasses, mosses, and a few tough plants like daisies and dandelions, as well as worms, beetles, ants, and many other insects. A south-facing, sunny flowerbed will be the home of many different flowering plants, plus insects such as ladybirds and greenfly, and worms and beetles in the soil. If there is a garden pond, it will be home to water beetles and dragonflies, frogs, fish, and slugs as well as water plants and pondweeds.

Each organism needs a different sort of habitat. Dragonflies could not survive in a hot, sunny flowerbed and greenfly do not live on grass.

WHAT IS YOUR NICHE?

Most habitats are home to a whole range of different living things, all of them taking up a different **niche**. Some organisms will be growing in the soil, some will feed at night and some in the daytime, some will eat plants and some will eat animals. But just what niches are available in the living world?

In any habitat there will be the **producers** (usually plants) making the food that everything else depends on. There will be tall and short producers, producers needing bright light and those that can manage in the shade. There will be the animals that eat plants (herbivores) and animals that eat other animals (carnivores). Herbivores range from microscopic animals which eat microscopic plants, through grazing animals such as cows or bison, to elephants, the largest living land animals, which can knock down entire trees. Herbivores may eat nothing but bamboo, or only fruit and berries, or dig up roots to eat – but they always feed only on plants.

Carnivores range from large **predators** such as lions and tigers, which eat herbivores such as zebra and antelopes, through to tiny predators such as shrews and hydra which eat insects or microscopic animals. This group even includes **parasites** such as tapeworms which live inside the gut and feed from their host. Scavengers feed on the remains of carcasses left by predators.The **decomposers**, such as bacteria and fungi, rot down the bodies and the waste products of the animals and plants. So there are no shortages of niches for living things, whichever habitat they might live in. In any given place there is a whole **community** of different organisms living, interacting, and depending on each other.

The snail is often overlooked in a habitat – yet the part it plays by eating plants and being eaten by birds is just as important as the role played by any of the larger organisms in the habitat.

SCIENCE PIONEERS Victor Shelford

Victor Shelford (1877–1968), an American biologist, was one of the very first people to recognize the importance of the relationships between animals and plants in the places where they lived. In his early years he spent a lot of time studying the sand dunes of Indiana. He was fascinated by the way the animals and plants depended on each other in that difficult environment. He developed the idea of ecology as the study of communities of organisms. His ideas became widely read when they were published in his best-known book, *Animal Communities in Temperate America*.

Victor Shelford studied in the sand dunes of Indiana in the US, earning them the nickname "the birthplace of ecology".

CHAINS AND WEBS

The animals and plants in a habitat are linked together in relationships, which scientists describe as **food chains** and **food webs**. Food chains give us the simplest picture. They all begin with an organism that can make its own food – usually a plant. The plant is eaten by an animal, which in turn is eaten by another animal, and so on. For example, grass is eaten by cows, and cows are eaten by people. At the end of every chain are the decomposers (fungi and bacteria), which rot down the bodies and the waste. But in most habitats, the picture given by food chains is too simple. Herbivores eat a wide variety of different plants and carnivores have lots of different **prey**, so a more accurate picture of the links between living organisms in a habitat is a food web. Plants are known as producers, because they produce all the new food. Animals that eat plants are known as **primary consumers**, while animals that eat other animals are called **secondary consumers**, and so on.

A food web sets out to show all of the different feeding relationships between the organisms in a habitat.

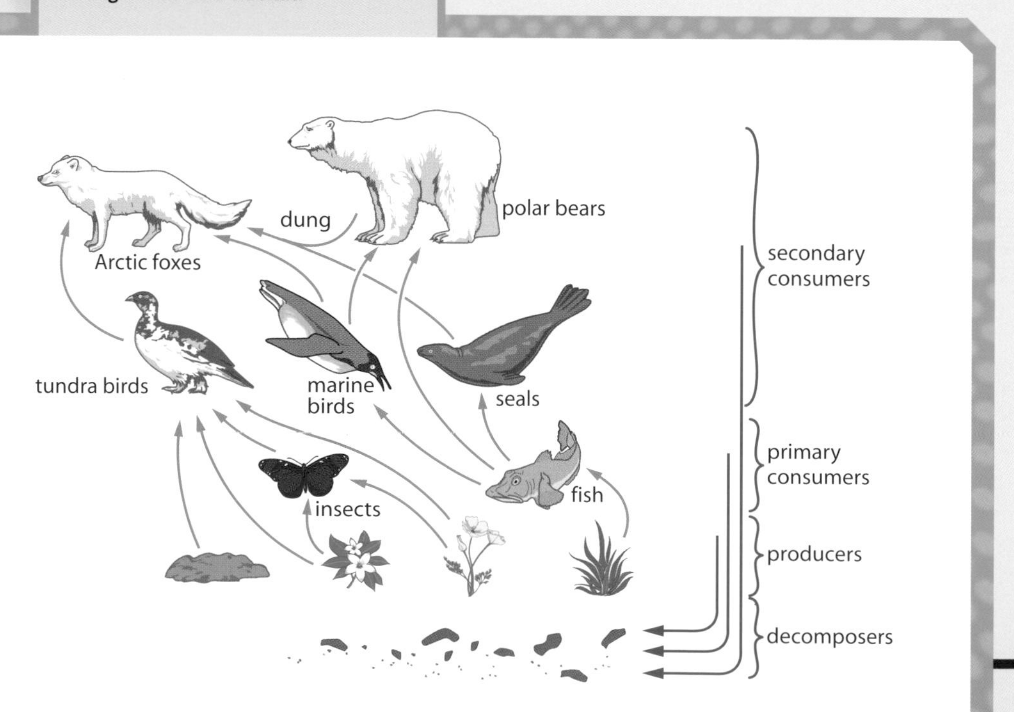

KEY EXPERIMENTS the work of Charles Elton

In the 1920s, Charles Elton, a young biologist at Oxford University, travelled to Bear Island off the northern coast of Norway. He wanted to see how the animals managed to survive this tough environment.

There are not many different types of plants on Bear Island so it was quite easy for Elton to see which animals fed on the plants – and on each other. Arctic foxes were the main large carnivores. During the warmer months they fed mainly on birds like ptarmigan and sandpipers, which only came for the summer. But in the winter, when even larger carnivores in the form of polar bears turned up on the island, the foxes fed on bits of dead seal left by the bears and even some polar bear dung! The birds in their turn fed on the leaves and berries of the **tundra** plants, or on plant-eating insects. Elton described the links between plants, insects, birds, and fox as a food chain. He said the first link of a food chain is a plant trapping energy from the Sun by photosynthesis.

But almost as soon as he had come up with the idea of food chains, Elton recognized that it was too simple. Each chain showed only one of the feeding relationships of each animal. So he built all his observations up into what he called a food web – a term we still use today.

Bear Island is a very stark environment. This is where Charles Elton first watched food chains and food webs in action.

HOME SWEET HOME

In order to survive, many organisms have changed or adapted to cope with the conditions where they live. Some adaptations are quite common – wings for flying, fins for swimming – while others are very unusual indeed. The most extreme and unusual adaptations are usually only found in very specific habitats. Before looking in more detail at some of the adaptations, it is worth looking at some of the environments.

HIGHEST AND LOWEST

In spite of their height, many of the world's mountain ranges are home to animals and plants. The biggest problem for animals at **altitude** is the lack of oxygen in the air as a result of very low **air pressure**. However, the weather conditions are often extreme as well. High winds, low temperatures, snow and ice, all make life difficult for plants and animals. Mount Everest, at 8,850 metres (29,035 feet), is the highest mountain in the world and little beyond bacteria can live at the summit. But the slopes of many other mountains are well-used habitats.

Mountains can be a tough place to live, but many different species of animals and plants still manage to call them home.

At the other extreme is the deepest part of the ocean – the Challenger Deep in the Marianas Trench, in the Pacific Ocean, east of the Philippines. This is 10,911 metres (35,797 feet) below sea level. In the ocean depths it is pitch black because no light can reach that far through the water. The pressure is enormous and it is very cold, usually about 2–3 °C (36–37 °F), yet some living things manage to survive.

FROM VERY HOT...

The hottest places to live are the deserts of Africa, Asia, Australia, and South America. However, the desert can also get very cold at night. The average daytime temperature for a desert is 30–38 °C (86–100 °F), but at night the temperature drops to around -3 or -4 °C (27–25 °F). The animals and plants that live in the desert have to be able to cope with extremes of temperature. Living in the tropics can be very hot too, at around 22–27 °C (72–81 °F). Here it stays hot all day, all night, and all year round!

...TO VERY COLD

Some parts of the world are extremely cold. The areas around the North and South poles are obviously chilly places to live, and the organisms that survive there have to be specially adapted to cope with the cold. But there are other very cold places – the 4,000 people who live in the village of Oymyakon in Siberia, Russia, cope with temperatures that fall to around -68 °C (-90 °F) in winter. Somehow plants and animals manage to survive here, too!

Did you know..?

The coldest temperature recorded anywhere on Earth so far is -89 °C (-128 °F) at Vostok in Antarctica on 21 July 1983.

Adaptations for life

Most living things need food and oxygen to survive. They get energy from their food by **respiration**, and they need to move, grow, and reproduce. But organisms live in all sorts of environments, so they often have adaptations to help them thrive in their surroundings. Some of these adaptations are very common, while others are extremely rare!

THE FOOD MAKERS

The leaves of plants must be one of the most amazing adaptations found in living things – yet they are often taken for granted. Leaves contain little packets (**chloroplasts**) of green pigment or colour (**chlorophyll**), which catches and traps light energy. Chloroplasts also contain all the **enzymes** needed to use this trapped energy to make sugars out of carbon dioxide and water. Veins carry water into the leaves, and also remove the sugar and move it all around the plant. Leaves have small pores (called stomata) that open and close to allow gases in and out. Photosynthesis not only makes food, but also produces oxygen. This means that leaves are vital to all life on Earth.

IT IS TOUGH TO BE A PLANT!

Leaves all do the same job, but they have to do it in a huge variety of different habitats. So over thousands of years plants have adapted their leaves to allow them to live in deserts and tundra, forests, and temperate climates – and they come in some very spectacular shapes and sizes. One of the hardest conditions for a plant is when it is very dry. In fact all plants are in a constant battle to make sure they do not lose more water through their leaves than they take in through their roots. Plants get dry for many reasons. In very hot conditions,

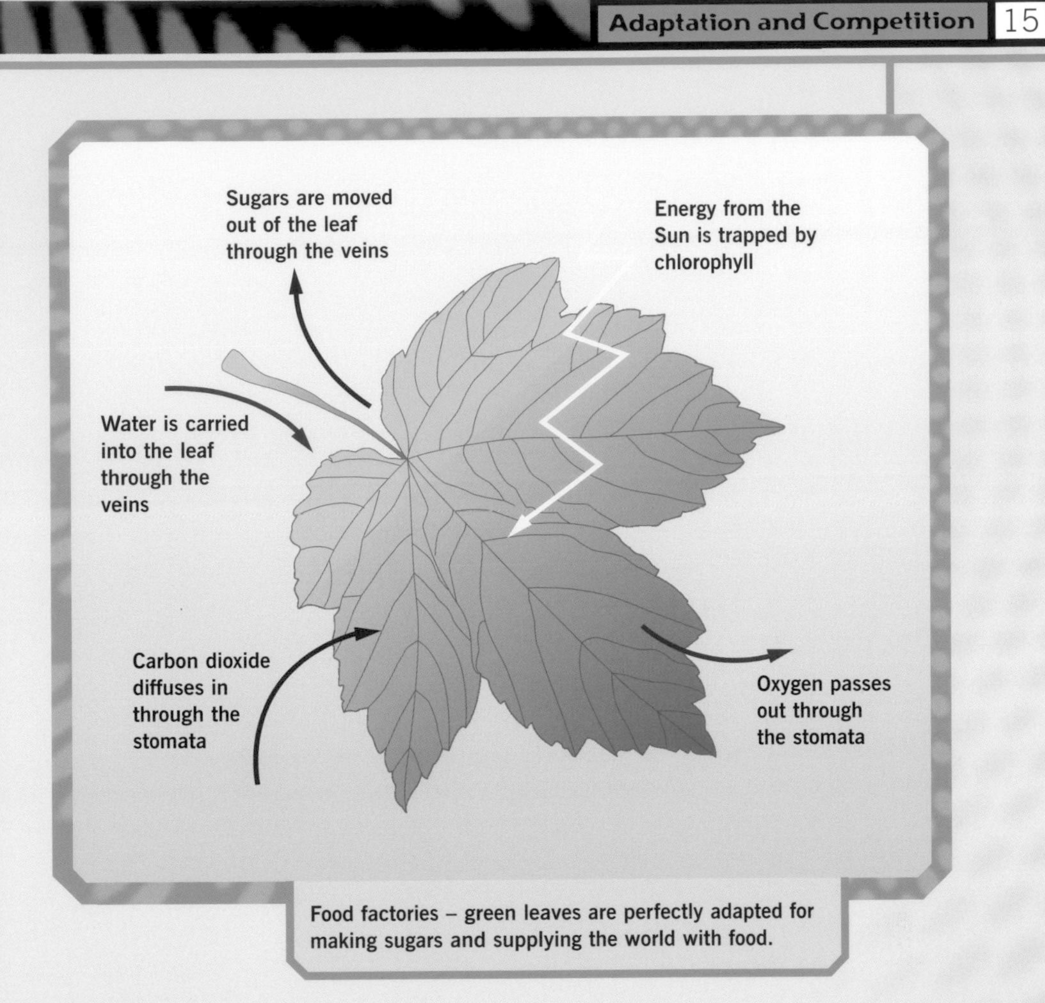

Food factories – green leaves are perfectly adapted for making sugars and supplying the world with food.

plants lose a lot of water through their leaves, while in very cold conditions the soil water is frozen and they are unable to take it up into their roots. In very salty conditions, water can be drawn out of the plant into the soil by **osmosis** instead of the other way round.

Did you know..?

Successful adaptations mean survival. One ancient bristlecone pine affectionately known as "Methuselah" in the White Mountains, California, is 4,767 years old, so it must be very well adapted indeed!

WATER WORRIES

Plants need water both for photosynthesis and to keep their tissues upright. If a plant does not get the water it needs, it eventually wilts, with drooping leaves and stems. So how do plants that live in dry conditions cope? Most plants try to prevent water escaping through their leaves by having a waxy layer on the upper surface of the leaf called the cuticle. Some plants trap a layer of moist air around their leaves by having hairy leaves, or curled leaves, or both. This stops the air around the leaf surface from moving, so it becomes saturated with moisture. This helps stop any more **evaporation** and means the plants need less water from the soil.

LIFE IN THE DESERT

Plants such as cacti reduce their leaves to tough spines which discourage animals from eating them. They then use their stems both to store water and to photosynthesise. Other plants, known as succulents, store water in fleshy leaves, stems, and roots when there is plenty about, and then use this stored water to survive in dry periods. These adaptations can be spectacularly successful. A mature apple tree loses the equivalent of a bath full of water from its leaves every day, but a large saguaro cactus loses less than one glass of water in the same amount of time!

Did you know..?

Some plants have adapted to become parasites. Parasitic plants invade other plants and grow using food and water that they steal from their host. Some of them give up on photosynthesis altogether and just live inside their hosts!

RECENT DEVELOPMENTS resurrection plants

Some of the most amazing plants in the world are the poikilohydric or resurrection plants, which can survive massive water loss. They have adapted not to prevent water loss but to cope with it when it happens. These plants can lose up to 95 percent of their water content without suffering permanent damage. When conditions get dry, the plants lose more and more water until all that is left are the small, shrivelled remains. The plant looks dead, but within about 24 hours of watering the tissues **rehydrate** and the plant looks as good as new! Dr Peter Scott and his team at the University of Sussex are trying to find out how this survival mechanism works. It looks as if the plants fill their leaves with sugars that protect them as they dry.

All over the world crops are lost and destroyed every year because conditions are too dry, and as a result millions of people do not get enough food. If scientists can find a way to produce "resurrection crops", then starvation might become a thing of the past.

This resurrection plant recovers with the addition of water. Scientists are trying to find a way of giving crop plants the same survival mechanism so they can survive when there is little rain.

WHERE IS THE NEXT MEAL COMING FROM?

Living things that cannot make their own food have to eat other living things to get the energy they need. Without food, animals die, so some of the most important adaptations in the living world are to do with getting and digesting food.

Eating plants might seem like an easy option. Plants do not move around and they do not – usually – fight back. But herbivores face some big problems when it comes to digesting their food. Plant cells are surrounded by a tough **cellulose** wall, which gives the plant support. Unfortunately most animals do not have the digestive enzyme (called cellulase) needed to break down cellulose so all the useful food material of the plant is locked away.

ADAPTATIONS FOR EATING PLANTS

Butterflies, moths, and bees are all insects that feed successfully on plants by avoiding cellulose altogether. Instead they feed on the sweet sugary nectar produced by the plant. Their special adaptation is a proboscis, a long, coiled mouthpart that can reach right down into the flowers to the nectar.

Aphids have developed sharp, pointed mouthparts that go through the cellulose cell walls and reach right into the plant's veins. The aphid can then suck up the sugary, food-rich fluid.

But some insects, such as termites, tackle the problem head on. Their favourite food is wood, which is mainly cellulose. Termites carry tiny little organisms called protists in their guts. Protists produce the enzyme cellulase, which breaks down the wood eaten by the termites into sugars. The protists get a protected home and regular food deliveries and the termites get their food broken down for them – good adaptation all round!

Lots of mammals are herbivores, but mammals cannot make cellulase at all. They use a number of different adaptations to cope with a plant-only diet. Many, such as elephants, horses, and cows, have large, flat, ridged back teeth and jaws that move from side to side as well as up and down. This makes sure plant material is thoroughly crushed and ground up,

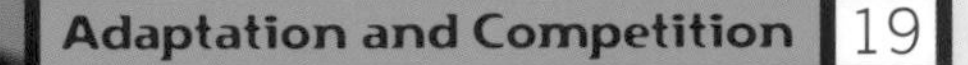

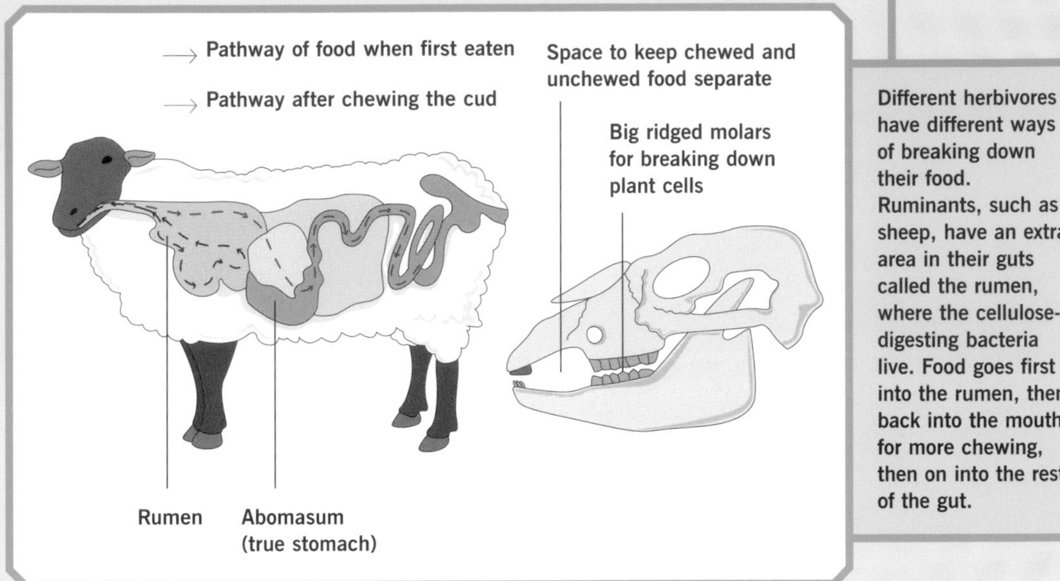

Different herbivores have different ways of breaking down their food. Ruminants, such as sheep, have an extra area in their guts called the rumen, where the cellulose-digesting bacteria live. Food goes first into the rumen, then back into the mouth for more chewing, then on into the rest of the gut.

releasing the digestible food from the cell walls. They also tend to have very long guts to help them digest every last bit of the food and ensure they get enough energy from it. Herbivores often rely on bacteria in their digestive systems to make the cellulase needed to break down the cellulose. Like the protists found in termites, the bacteria get a constant supply of food and a warm, protected environment, while the mammal gets sugar from the cellulose. Digesting cellulose is very difficult, so even with these adaptations most herbivores spend a lot of time eating and they produce large amounts of very bulky, cellulose-rich **faeces**.

Herbivores, such as this rabbit, need to spend a lot of their time eating plants in order to get all the food they need.

THE MEAT EATERS

Not all animals eat plants. Carnivores feed on other animals. Animals are made up of lots of protein (particularly their muscles and skin), with some fat and carbohydrate – all of which is quite easy to break down. Also, carnivores' food is high in energy, so carnivore guts are usually much shorter, and they produce relatively little waste material. They do not need to spend as much time feeding and digesting food as herbivores, but there are still difficulties to overcome.

RECENT DEVELOPMENTS the fastest predator in the world?

It takes a human driver about 650 milliseconds from seeing a red light to braking. But the star-nosed mole takes only 230 milliseconds from the moment it first touches its prey to gulping it down – faster than the human eye can see. What makes this even more amazing is that star-nosed moles live underground and are almost totally blind! Their main sense organ is a crown of fleshy tendrils around the nose, incredibly sensitive to touch and smell. It seems likely that they react so quickly because they cannot see what is going on. They need to grab the prey as soon as possible after touching it before it moves away or tries to avoid them. Kenneth Catania and his team at Vanderbilt University in Nashville, Tennessee have been studying these amazing moles to find out more about how quickly the brain can process information.

The ultra-sensitive tendrils of this star-nosed mole can try out thirteen possible targets every second.

This orchid mantis looks like a beautiful flower, but any butterfly that approaches will soon find itself on the menu for dinner! The camouflage that helps the mantises to catch prey also helps them to avoid being eaten by birds. One adaptation benefits the mantis in two ways.

FIRST, CATCH YOUR PREY...

Many carnivores are predators, which means they catch and kill other animals, their prey. But most prey animals do everything they can to avoid being eaten! So predators have adaptations that help them catch their prey, hold on to it, and kill it. Sharp claws and teeth, the ability to run fast, lots of stamina to keep tracking prey, poisons, stings, and the ability to hunt in teams are all important for some carnivores. Some rely on camouflage to hide until their prey get near enough to pounce on, while others lure their prey to them.

Not all carnivores kill their food. Some are scavengers. They feed on dead bodies and the remains of other animals' meals. Some carnivores are parasites and feed off other living animals – either feeding on their blood like fleas, or living inside their body like tapeworms.

SO WHERE DO YOU LIVE?

You can often work out where an animal lives by looking at it. Animals that live in very hot countries often have large ears, thin fur, and loose skin to help keep them cool. Animals that live in cold places tend to be the opposite, with short ears, thick fur, and fat layers under the skin to help prevent them from losing heat. When organisms live in more extreme conditions, they need some unusual adaptations.

DIVING DEEP

Seals and whales breathe air but have to dive under the water to find their food. Diving mammals like these need very special adaptations. They have very big lungs to take in lots of air so they do not need to breathe for a long time. They have extra blood to carry lots of oxygen and when they dive, the blood flow to non-essential bits of their bodies (such as the gut and the skin) is shut off. This means that blood is directed to the brain and the swimming muscles where oxygen is most needed. The heart rate slows to just a few beats per minute so less oxygen is used up. Finally, their bodies can cope with a build up of waste products (such as carbon dioxide) until they return to the surface. All these adaptations allow them to dive for 20 minutes or more at a time.

LIVING HIGH

Thousands of feet above the sea, the amount of oxygen in the air is much lower than at sea level – yet people still live in the mountains. Their bodies are specially adapted to cope. They have larger lungs, with many more tiny blood vessels and red blood cells than other people. This helps them get as much oxygen from the air as possible. Anyone who is not born in these conditions risks **altitude sickness** and even death when they climb to great heights.

KEY EXPERIMENT figs and fig wasps

There are about 700 different species of fig trees and each one has its own species of pollinating wasps, without which it would die. The fig flowers of the trees are specially adapted so that they attract the correct wasps. If a fig tree cannot attract the right species of wasp, it will never be able to reproduce. In fact, in some areas the trees are in danger of extinction because the wasp populations are being wiped out. Dr James Cook and his team at Imperial College, London are looking at the adaptations of the different wasps and their **DNA**, to try to work out the relationships between all the different species.

Female fig wasps have specially shaped heads for getting into fig flowers, and **ovipositors**, which place the wasp's eggs deep in the flowers. The male fig wasps vary. Some species can fly, while others are adapted to live in a fig fruit until they fertilize a female wasp. They then dig an escape tunnel for the female through the fruit and die!

This is a picture from Dr James Cook's team, showing fig wasps inside a fruit.

HOW DO ADAPTATIONS COME ABOUT?

Hippopotamuses are wonderfully adapted to life in African rivers and swamplands. However, they do not look much like their relations, the horses, which are adapted for spotting predators and running away fast on open plains. How do different adaptations arise?

All the characteristics of living organisms are controlled by their **genes**. This is the genetic information carried in the **nucleus** of every cell. Sometimes genes can change slightly during reproduction – there is a **mutation** – and this can result

The stripes of a zebra are clearly an adaptation to something – but what? Scientists have come up with lots of theories. Perhaps they help the zebra blend into the long grass, or it could be that the stripes confuse predators and make individual zebra harder to pick out. Another theory is that the stripes make zebras less open to attack by the tsetse fly, which carries sleeping sickness. No-one is exactly sure.

in a change in the offspring. The change could be anything from a longer neck to more chloroplasts in a leaf. If this new feature gives the animal or plant an advantage, such as making it more attractive to a mate or helping it catch food, it may well be passed on to the next generation. There are always more offspring produced than the environment can support, so the offspring best suited to life in a particular habitat will be most likely to stay alive and breed successfully. This is known as natural selection, and it has given rise to all of the different species we see around us today.

SCIENCE PIONEERS Darwin and Wallace

When Charles Darwin travelled on *HMS Beagle* to South America and the Galapagos Islands, he came across many organisms that showed clear adaptations to their environments and foods. He discovered that black marine iguanas stored up body heat so they could go into the cold sea and feed off the weed on submerged rocks. He also realized that finches had different shaped beaks depending on whether they ate seeds or nuts. Darwin made notes on all his observations and returned to England to write up his ideas about natural selection.

Alfred Wallace also travelled – first to South America and then to Borneo – and just like Darwin he was struck by the way each species was adapted to a particular way of life. He came to the same conclusions; that the animals and plants with the most useful adaptations would survive and breed. He wrote to Darwin about his ideas, causing Darwin to worry that Wallace would publish a book of these ideas before he could! In the end, they both put papers forward at the same time, but it was Darwin who published *On the Origin of Species by means of Natural Selection*. His book changed the way we look at adaptations, and biology itself, forever.

A changing habitat

Some organisms do not like change in their environment. They live in habitats that are very stable. The oceans of the world are home to countless millions of plankton and fish, and apart from the top few metres, the temperature of the water barely changes from one year to the next. Tropical rainforests, which have more species living in them than any other habitat on the land, are also very stable. They show little variation in temperature, rainfall, or even the day length – there are about 12 hours of daylight and 12 hours of darkness every day of the year.

Stability like this is relatively rare. Many organisms live and thrive in habitats that change dramatically. Some may change seasonally with big temperature differences between winter and summer. Others habitats, like the desert, vary greatly between day and night, while organisms that live in tidal zones have to deal with completely different conditions four times a day.

THE CHANGING SEASONS

In many parts of the world, there are big variations in temperature between summer and winter. In the UK, the average summer temperature is around 20 °C (68 °F) with a winter average of only 5 °C (41 °F). This seems quite a big change until you realize that in Alaska the swing can be from 39 °C (102 °F) on a summer's day down to -17 °C (1 °F) on a winter night. The amount of daylight changes dramatically too, from about 16 hours of daylight in the UK summertime to 8 hours in the winter, or from 24 hours daylight during summer in the Arctic Circle to none at all during the winter! The animals and plants that live in these areas have to be able to survive both extremes of the year.

Many organisms do not even try to cope with the extremes of their environment. They **hibernate** or become **dormant** during unfavourable conditions, an adaptation in itself. Many plants grow, flower, scatter their seeds, and die before the harsh conditions begin. They simply leave their tough seeds to survive the winter and **germinate** next spring. Many trees cannot cope with winter conditions either. In the short days of winter there is not enough sunlight for the leaves to make much food by photosynthesis, and they would all be killed by the frost. So the trees become dormant. They lose their leaves and slow down all the processes of life for the cold, dark months, before surging back to life in spring.

The dramatic colours seen in these trees show us that they are preparing to shut down for winter.

Did you know..?

Many conifers keep their leaves all winter by having several different adaptations. Their leaves are needles – thin, tough with a very thick waxy outer layer, and with stomata that can close very tightly so the leaves lose very little water. Their cell chemistry helps them cope with very cold conditions, and the typical shape of a conifer means that snow slides off the branches rather than breaking them.

ANIMALS IN WINTER

Winter not only means cold weather and short days for animals, it often brings food shortages as well. Plants are in short supply in the winter, so many animals have very little food. To deal with these problems, some animals change their behaviour, some change the way they look, and others change the way their bodies work.

Many animals grow thick, shaggy coats, which keep them warm through the winter months. Many also build up thick layers of body fat over the summer months. This acts as an **insulating** layer to help keep them warm in winter and gives them an energy store for when food is in short supply.

Some animals, such as dormice and ground squirrels, take fat storage a step further. They build up really thick layers of fat, and then hibernate through the worst of the winter. They usually build a cosy den or burrow, and then allow their bodies to slow down. Their heart and breathing rates drop dramatically and their body temperature falls to the same level as the surroundings. This saves huge amounts of energy and lets the hibernating animals survive for months.

Some animals, such as the ptarmigan, change the colour as well as the thickness of their coats in winter. White feathers mean the ptarmigan is less likely to be spotted by predators than if it kept its brown summer feathers.

RECENT DEVELOPMENTS biological antifreeze and genetic engineering

Fish are cold-blooded, which means that normally their bodies are at the same temperature as their surroundings. For fish that are swimming in water that is around freezing point, this could mean that their tissues freeze and they die. But some have an amazing adaptation. They make a chemical that is very similar to the antifreeze you put into a car in the wintertime – and it does exactly the same job! Fish with antifreeze, such as the cold-water flounder, have tissues that do not freeze at 0 °C (32 °F) so they can survive the coldest winters. Now scientists have identified the gene that produces the antifreeze **molecule** and moved it into other genetically modified organisms. They have produced frost-resistant tomato plants, which make their own antifreeze to protect the crop.

Sometimes, genetic modification does not produce the change you wanted. The antifreeze gene was also introduced to salmon so that they could be farmed in colder waters. But the salmon did not make antifreeze. The new gene had a completely different effect. The genetically modified salmon grow several hundred times faster than normal fish, an adaptation that helps salmon farmers but not the salmon!

Did you know..?

When an American woodchuck hibernates, its heart rate drops from the normal 80 beats a minute to 4 or 5 beats a minute and its body temperature drops from 37 °C (99 °F) to 3 °C (37 °F).

A LONG WAY HOME

Some animals cope with the changes by simply avoiding them altogether. Some types of birds, fish, insects, and mammals can travel hundreds and even thousands of kilometres to avoid harsh winter conditions, or to get away from summers that are too hot. This is called migration. Some birds, such as swallows and martins, visit the UK to breed in the summer and then fly back south to Africa for the winter months. Across the Atlantic, snow geese breed in the brief Arctic spring and summer of Alaska and then fly across the US to spend the winter in the relative warmth of Mississippi, Texas, Louisiana, or even Mexico.

Migration can take place over land as well as by air. Whales, fish, and mammals such as wildebeest are just some of the animals that travel huge distances to reach the food and conditions they need. The adaptations that make these migrations possible are still not fully understood. The animals need to navigate, and they need stamina as well. Senses that detect the magnetic poles of the Earth, and an ability to navigate using the Sun, Moon, and stars are both possible adaptations in the great migrators.

In winter, monarch butterflies from the northern USA fly in huge numbers to the Gulf of Mexico to breed, flying north again in the spring – an amazing journey for such small and delicate animals.

KEY EXPERIMENTS Eberhard Gwinner and his willow warblers

In the 1980s, Eberhard Gwinner carried out some famous experiments on willow warblers. These small songbirds usually fly from Germany to Spain in the autumn and then carry on until they reach South Africa. They spend the winter in the African sun before flying back to Germany for the summer. Gwinner carefully controlled the conditions in which he kept his birds, with artificial day and night for their whole lives. Yet they still became very active as autumn approached, and spent a month facing mainly southwest – the direction of Spain. They then changed direction to face southeast, the direction they needed to fly in to get from Spain to South Africa. They stayed active facing in this direction for another three months – the time it would take them to make the flight. This pattern of behaviour was **innate**, which means it was not behaviour they had learnt from their parents.

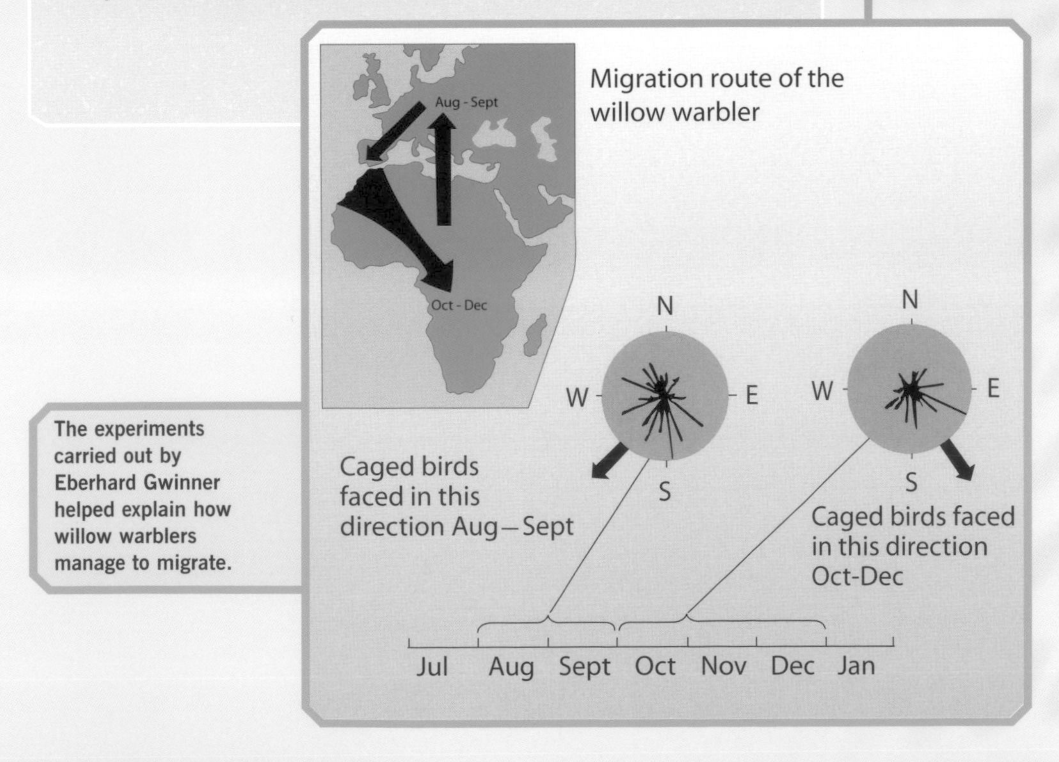

The experiments carried out by Eberhard Gwinner helped explain how willow warblers manage to migrate.

GETTING THROUGH THE DAY

It is not only the seasons that bring big changes. In many desert habitats the average daily temperature is around 38 °C (100 °F) while the night time temperature can fall to below zero. Many desert animals have adapted their behaviour to help them avoid the problems this can cause. Tortoises, desert rats, snakes, scorpions, and many other animals spend the hottest parts of the day hidden under rocks or bushes, in holes or in burrows. Keeping in the shade – or even better, underground – helps them avoid overheating. These same hiding places keep them warm at night, because below the surface or hidden in a crevice the temperatures do not change so much.

Camels also have specially adapted eyes, noses, and ears to prevent too much sand blowing into them. Camels are perfectly adapted for life in the desert!

THE DESERT WAY OF LIFE

Camels are the biggest desert mammals and need many adaptations to cope. They have very efficient kidneys, which means they produce very little **urine**. This helps them to cope with a lack of water. Most mammals keep their bodies at a constant temperature. Camels do not sweat, which again saves water, but this also means they do not cool down. During the day a camel's body temperature may be as high as 40 °C (104 °F) but it falls to around 34 °C (93 °F) in the night. Camels can survive like this because most of their cells are very temperature tolerant.

Camels are also adapted to the desert in other ways. They have amazingly large feet so they can walk on the sand with no problem as their weight is spread out over a large area. They have thick coats, which keep the heat of the Sun out during the day and the body warmth of the camel in at night.

THE COLD-BLOODED PROBLEM

Desert insects and reptiles are cold-blooded so they rely on their surroundings for heat. In the cold desert nights they retreat to holes and cracks in rocks, but they still get very cold. As a result the chemical reactions of their bodies slow right down and they cannot move very fast. When the Sun rises and the desert begins to warm up, the cold-blooded animals come out – slowly! They spread themselves out on rocks and the ground to absorb as much energy from the Sun as possible. Once they are warm enough, they will move around and feed, until the day becomes too hot, when once again they hide away until the worst of the heat is over. Larger reptiles may open their mouths and gape in the heat of the day so that water evaporates from their tissues and cools them down. Then it is a case of balancing the need for water against the need to cool down.

So desert animals, just like the desert plants we looked at in chapter 3, must have special adaptations that allow them to cope with the harsh conditions of their habitat.

HIGH TIDE, LOW TIDE

The animals that live between the high and low watermarks of the seashore probably have to deal with sharper contrasts in their environment over a short time than any other creatures on Earth.

When the tide is in, seashore animals and plants must be able to cope with being covered by salty water, and be able to withstand the crashing power of the waves. When the tide goes out, it is the drying heat of the Sun and wind that are the problems – and often a complete lack of water.

Some organisms have body adaptations to deal with the conditions, some adapt their behaviour to avoid the problems – and some do both. Seaweeds cannot move about, so most of them have a tough, strong root called a holdfast, which keeps them fastened tightly to the rocks. Some of them also have bladders filled with a watery fluid or a coating of slime to help them cope with the drying effects of the Sun.

Organisms in rock pools have to cope with the waves as the tide comes in. When the tide is out, they also have to deal with the water evaporating steadily causing the salt concentration of the pool to rise.

SEASHORE ANIMALS

Limpets and barnacles are wonderfully adapted for seashore life. They are **molluscs** with a soft, muscular body completely protected by a tough shell. They live on rocks and when the tide goes out the shell is clamped down hard over the animal, trapping water inside and preventing it from drying out. It is extremely difficult to remove them from the rock, which is an added protection both from the action of waves and from the sea birds looking for food when the tide is out.

Many animals such as sea urchins, sea anemones, starfish, mussels, shrimps, and crabs hide in the cracks and crevices of rocks, in rock pools, and under seaweed to avoid the Sun and predatory birds.

Not all beaches have rocks. On sandy beaches there are few places to hide when the tide goes out. The only way out is down. So the worms, shellfish, and crabs that live on sandy beaches tend to have adaptations that help them to dig downwards into the sand as the tide goes out. This allows them to remain cool and safely hidden until the tide comes in again.

Every November, a spectacular carpet of about 60 million red land crabs scuttle across Christmas Island from the forest to the ocean to mate and lay eggs. They cross roads, farms and everything in their path on the way to the sea.

Did you know..?

Some seashore animals have become so well adapted to life out of water that they move away from the sea completely for most of their lives. The red land crabs of Christmas Island spend most of their lives in the rainforests at one end of the island, but they still need the sea for breeding and return to the seashore once a year.

The role of competition

Adaptations are important in helping organisms to survive in their habitat. But an animal or plant does not grow on its own – it grows alongside lots of other living things. As there is only a limited amount of resources, food, water, and places to shelter, there is often competition between living organisms for the things they need. The winners of the competition will get the most food, water, or shelter, making them the most likely to survive.

Organisms need adaptations that will let them compete successfully against other members of their own species and against different species. The organisms that are best adapted are most likely to breed and produce healthy offspring. So competition affects the size of populations and as a result the whole natural world.

The biggest cause of competition is the drive of all living things to reproduce. Animals and plants need to get as much food as possible, so they can grow as quickly as possible, so they can reproduce as soon and as often as possible. It is not just having lots of offspring that matters – it is making sure that as many of them as possible survive to grow up and breed themselves.

Did you know..?

If a single bacterium was given perfect conditions, and so were all of its offspring, then after 48 hours there would be 22,000,000,000,000,000,000,000,000,000,000,000,000,000,000 bacteria. These would have 4,000 times the mass of the Earth!

KEY EXPERIMENT growth in a perfect world

If bacteria are provided with everything they need – plenty of food, oxygen, and warmth – they will grow and divide every 20 minutes. This means the population grows at an amazing rate! Fortunately this does not usually happen because as the number of bacteria increase, they begin to compete with each other for food and oxygen. The rate of growth slows down, but there are still more and more bacteria building up. They start to poison each other with their waste products and some of the bacteria begin to die. For a while the numbers of bacteria dying balances the new bacteria being produced, but eventually the food and oxygen levels get so low – and the poisonous waste levels get so high – that far more bacteria are dying than are managing to reproduce. The numbers of bacteria in the population fall and eventually, when all the food and oxygen has gone, the whole population dies. Although bacteria reproduce very differently to most animals and plants, they can still show us just how important competition between individuals is in controlling the numbers in a population.

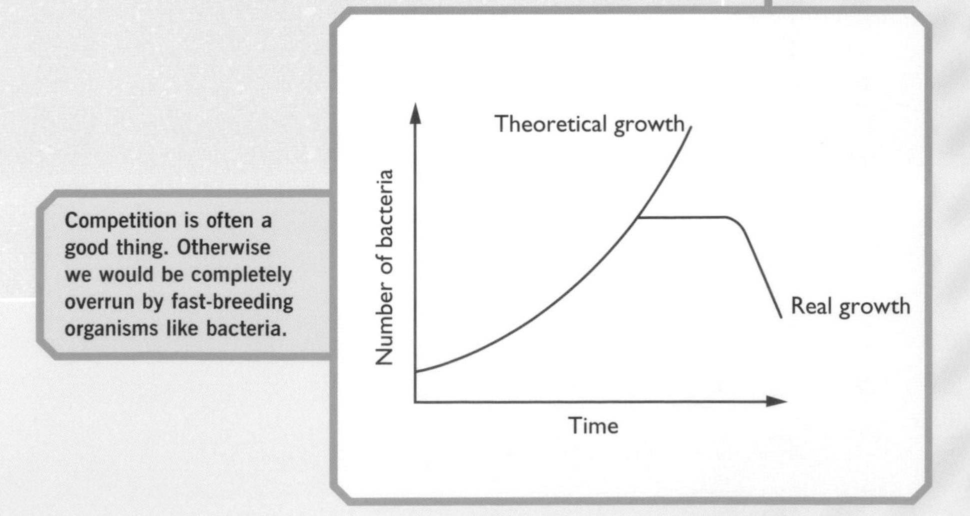

Competition is often a good thing. Otherwise we would be completely overrun by fast-breeding organisms like bacteria.

COMPETITION IN PLANTS

Plants might look like peaceful organisms, but the world of plants is full of cut-throat competition. Plants compete with each other for light, water, and for minerals from the soil. When seeds from different plants land on the soil and start to grow, the plants that grow fastest will win the competition against the slower-growing plants.

If a plant sheds its seeds and they land nearby, the parent plant will be in direct competition with its own seedlings. The parent plant is large and settled, so it will take the largest share of water, minerals, and light, depriving its own offspring of everything they need to grow successfully. If the seeds all land close together – even if they are a long way from their parent – they will compete with each other as they grow. To prevent problems like this, many plants have special adaptations (called dispersal mechanisms) that help them to spread their seeds over as wide an area as possible.

SPREADING THE SEEDS

Seeds come in an enormous range of sizes, from the tiny seeds of the rattlesnake plantain, which only weigh 0.000002 gram (0.00000007 ounce) to the giant seeds of the coconut palm, weighing over 27 kilograms (60 pounds)!

Many plants use the wind to help them spread their seeds. Some produce seeds so small that they are carried easily by air currents. Many others produce fruits with special adaptations

Tumbleweeds, found on the plains and deserts of Northern America, use the whole plant to scatter their seeds. They grow into a ball shape, flower, and produce seeds. When the seeds are ripe, the plants break off at the roots and are blown away, travelling miles across the plains and scattering seeds as they go.

that help their seeds to travel as far from home as possible. The fluffy parachutes of the dandelion clock and the winged seeds of trees such as the sycamore and the lime are common examples of flying fruits.

Some plants use mini-explosions to spread their seeds – the pods dry out, twist, and pop, flinging the seeds out and away. Some fruits rely on water to carry them. Coconuts will float for weeks or even months on ocean currents that can carry them hundreds of kilometres from their parents!

HITCHING A RIDE

A wide variety of plants depend on animals to scatter their seeds for them. All over the world, plants produce juicy berries, fruits, and nuts to tempt animals into eating them. Once the fruit gets into the animal's gut, the tough seeds simply travel through and are deposited with the waste material in their own little pile of fertilizer! By this time the animal may well have travelled far from the place where the fruit was eaten. Birds are used most commonly to spread seeds, but many types of herbivores are also involved. There are even fruits that are sticky or covered in hooks, which get caught up in the fur or feathers of a passing animal. The fruits are carried around until they fall off or are removed by grooming hours or even days later.

Banksia aemula are adapted to release their seeds after a bush fire. They then quickly grow back in the burnt area.

Did you know..?

Bush fires are very common in Australia, and they destroy many plants and animals. But the fruits of the plant *Banksia aemula* are so hard and woody that they will only open and release the seeds once they have been burnt in a fire. This adaptation makes sure that not only do the plants survive bush fires, but also that they get a head start on everything else once the fire is over.

COMPETITION IN ANIMALS

Animals compete for food, water, space to live, mates, and breeding sites just as fiercely as any plants. Competition for food is particularly common. Herbivores eat plants and often several different species will all eat the same sort of plants. The animals that eat a wide range of plants are more likely to be successful because they have a greater choice of possible food.

In the same way, carnivores compete for prey with their own species as well as other carnivores. Again, the animals that eat the biggest range of prey will be the most successful. To reduce the problems caused by competition, many animals set up territories and defend them against other animals to try to make sure they have enough food for themselves, a mate, and some offspring.

It may not be much – but it is home! These cape gannets in South Africa each defend just enough territory to build themselves a nest. If they cannot find a territory, they cannot reproduce.

RECENT DEVELOPMENTS ant invasion!

In California, native ants are losing territory to a vast supercolony of tiny Argentinian ants. The invaders are not winning the competition for California's ant habitat by fighting, as they are not very aggressive. But they are extremely good at sensing and finding food so there is not enough food left for the native ants.

The takeover by Argentinian ants in California is having a big impact on the local population of horned lizards. These lizards live on a diet of Californian ants, but they do not feed as well on the new species. Their numbers have fallen by almost 50 percent in the areas where the Argentinian ants have taken over the territory.

Territories may be large – several hundred square kilometres for some of the bigger carnivores – or very small. Some birds that nest in huge colonies have barely a square metre to call their own. Birds often mark their territories with song, while other animals use markers like piles of droppings, scratches on trees, urine spraying, or scent rubbings.

Animals will check their borders regularly and often attack any intruders who seem to be a threat. A single animal or a pair usually defends a territory, but in some cases a pack of animals like hyenas, or a vast colony of animals like ants, will defend a territory together.

FINDING A MATE...

Some of the most spectacular adaptations in the animal kingdom are linked to finding a mate and then successfully rearing the offspring. These adaptations can be physical – the antlers of deer, the colours of the belly of a male stickleback – or affect the ways an animal behaves. Nest building, courtship displays, and the bonding between parents and their babies are all adaptations that help animals raise their young.

In many species, the male animal puts a lot of effort into impressing the females, who want the best father possible for their offspring. Many male mammals display to the female and they may fight with other males to get her attention. Some birds have spectacular adaptations to help them stand out. Male peacocks and birds of paradise have the most amazingly beautiful feathers, which they use for displaying to other males (to warn them off) and to females (to attract them).

The male peacock puts on a dramatic display to attract his mate and his spectacular feathers demand attention. However, the females of this species are much less glamorous. Their drab, brownish feathers are a different adaptation – one that makes them hard to spot as they sit and look after their eggs.

...AND RAISING THE KIDS!

Once the young arrive, many more adaptations come into play. Young chicks gape their beaks to show brightly coloured mouths, and the parent birds stuff food in. The fur of young deer is patterned to give them camouflage, and they will lie quietly for hours on end until their mother returns to feed them. Mammals have some of the most impressive adaptations for successful breeding. The **embryo** grows safely within the body of the mother, and after birth, the mother produces milk, guaranteeing a food supply for her young ones.

One of the most important adaptations for survival in most animals is the instinctive behaviour they are born with. Instinctive behaviour is behaviour that does not need to be learned. Most **invertebrates** have short lives, so they need instincts, as they do not have time to learn everything. **Vertebrates** often live longer, but instincts are particularly important for helping very young animals survive until learning and experience can take over.

SCIENCE PIONEERS Niko Tinbergen and the importance of instinct

Niko Tinbergen showed that each piece of instinctive behaviour has a simple trigger that sets it off. So when a herring gull parent returns to the nest with food, it points its beak downwards and waves it from side to side in front of the chicks, showing a red spot marking. The chicks then peck at the beak and the parent feeds them. Tinbergen found that a pencil with three red spots moving from side to side would make herring gull chicks peck just as furiously as at their parent's beak. The vertical shape with red dots moving from side to side triggered the instinct to peck, not mum or dad!

Nature red in tooth and claw

The natural world looks calm and tranquil as we see it from the windows of our homes and cars, or take a walk in the countryside. In fact, it is the site of life-and-death struggles, as living organisms compete with each other for survival. Bacteria infect plants and animals alike, fungi feed on dead and living tissue, insects battle to the death, billions of plants are torn and shredded every day, while carnivores prey on the young, the weak, and the old. This is why the poet Alfred Lord Tennyson described nature as "red in tooth and claw". The competition for survival is not just between individual organisms, but also between different populations living in a community.

THE COMMUNITY EFFECT

In any community, different groups of animals and plants will use similar resources in a habitat. Food chains and webs become woven together and all the populations in a community end up interacting in some way. What happens to one population has a knock-on effect on many others in the area. Usually the interactions work to balance out any changes, so that although populations go up and down, the overall balance of organisms stays the same.

For example, if an area has a spell of warm weather with plenty of rain in early spring, the plants will grow particularly well. This will mean that there is lots of food for herbivores such as mice and rabbits, so they will breed successfully and raise lots of young. This provides food for predators, and animals such as foxes and owls will be able to feed their babies, so the population of predators will grow. During that year, all the populations will do well, but the big herbivore population means that eventually the plants get overgrazed.

With fewer plants now around, but still lots of herbivores, some animals die of starvation. Also, as the young predators grow, more of the herbivores get killed and eaten by the predators, so suddenly the numbers of the herbivores will be falling fast. This in turn affects the predators. Fewer mice and rabbits mean some of the young foxes and owls die of hunger or move off into new territories. Now all the populations start to fall in numbers. But this can start the cycle off again. A fall in the number of carnivores will mean more herbivores can now survive and, as the plants begin to recover, the herbivore population will begin to grow again. This picture of good times followed by shortages is a very simple model of how the balance of nature works.

Predator and prey populations – like the fox and grouse – tend to be linked together in cycles like this. After prey numbers increase, predator numbers increase as there is more food available. The number of prey then falls as predators eat them. Predator numbers eventually fall as less food is available. The whole cycle then starts again!

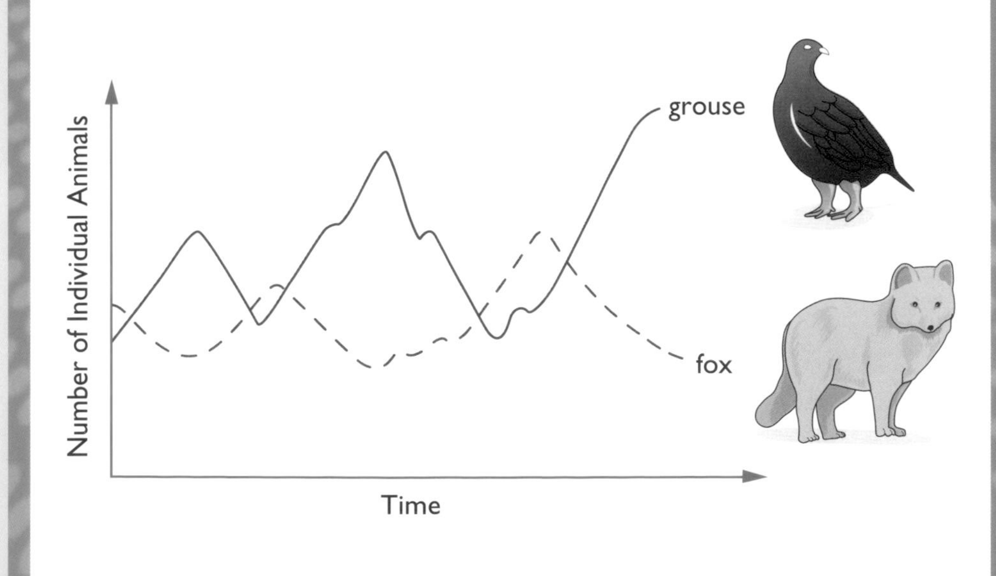

REAL-LIFE STORIES

Sometimes population numbers drop for no obvious reason. Scientists are now finding that hidden problems can have a big impact on the numbers of animals in a population. Parasites, for example, are not easily seen from the outside, but if animals are badly infected by parasites, they may become easy prey or may just die and this can have a big effect on a population.

KEY EXPERIMENT lynx and hares in Hudson Bay

For many years from the early 1800s, North American fur trappers sold the pelts of snowshoe hares and lynxes to the Hudson's Bay Company in Canada. The company kept a record of all the furs, and scientists used them to look at the populations of these animals over many years. The hare populations build up to a peak and then crash every nine to ten years. The lynx population does exactly the same thing, with a slight delay. Scientists discovered that the changes in the lynx population are the result of the changes in the numbers of hares. However, they then discovered that even when snowshoe hares live in areas where there are no lynxes to prey on them, they still show a similar ten-year cycle. The crashes in the hare population are the result of peaks and crashes in the populations of the plants they eat, which in turn are affected by the weather, and by insect pests.

Snowshoe hares have many adaptations, including their "snowshoe" paws. But they cannot avoid the boom and bust pattern of their population growth because they have no control over how much plant food actually grows.

The relationships between living organisms are rarely simple. Sometimes a change in one type of organism has a big effect on others, but sometimes it does not. There are many different plants in a habitat, and many different animals feeding on the plants or on each other. Sometimes, if something happens to one plant or one type of animal, the other animals simply eat more of other things, so the change makes very little difference. For example, Arctic foxes feed largely on lemmings because there are usually lots of lemmings and they are quite easy to catch. But every few years the lemming population gets too big. They migrate, many of them die, and the population crashes. This has a small effect on the numbers of Arctic foxes, but not as much as might be expected. This is because the foxes simply start eating more of the geese that are part of the same community.

The regular crash in the lemming population does not have a major effect on the number of foxes because they simply swap prey!

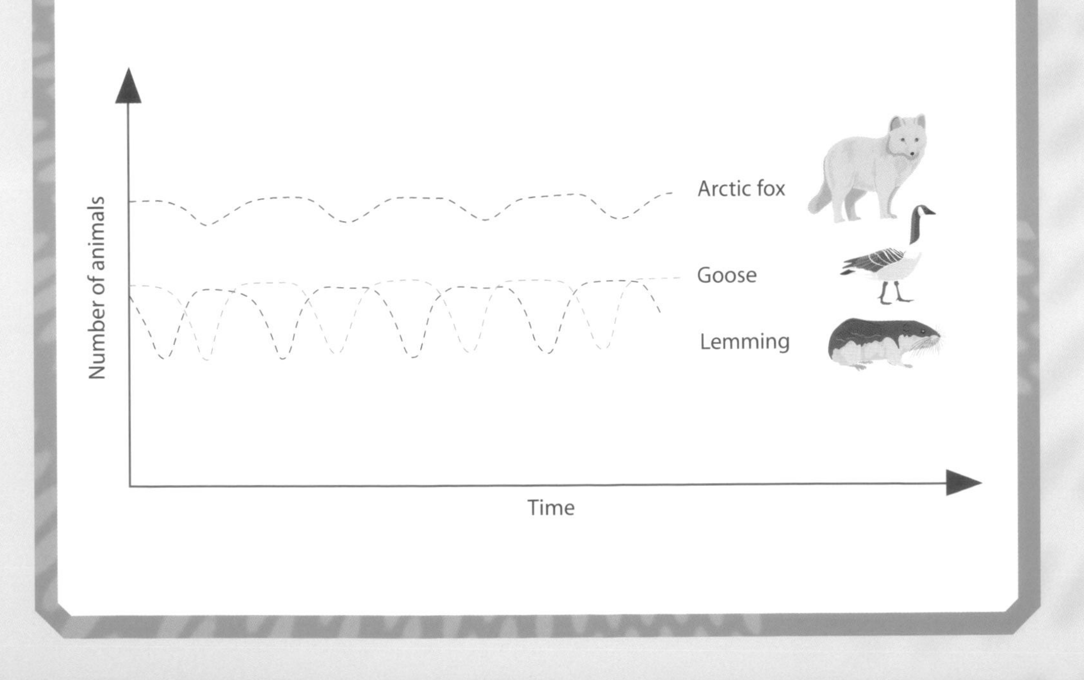

AVOIDING COMPETITION

In most environments in the living world, animals and plants live together in communities and the relationships between them work well. These may be based on competition for food or mates. Animals may be hunting each other every day, but over months and years the numbers of organisms follow a regular pattern and the community thrives. This is the result of adaptations that give the plants and animals slightly different needs, allowing them to live side by side. The best way to win a competition is to avoid it altogether.

If there are a number of plant-eating animals in a particular habitat, they might all be competing with each other. But if one type of animal eats grasses, whilst another browses on bushes and trees, another eats mainly seeds and berries, and a fourth feeds mainly on tough weed species, there will be little competition. Similarly with carnivores, when some eat insects while others eat worms and slugs, and some eat rabbits while others prey on small birds, direct competition for food is kept at a minimum. In most habitats, when you look at exactly what the different animals feed on, it becomes clear that they avoid competition with each other as far as possible. Even plants manage to steer clear of direct competition – some grow well only in bright light while others are adapted for shade, some are adapted to grow on dry soil while others prefer wet areas.

Did you know..?

Different types of African dung beetles avoid competition with each other by attacking dung piles at different times of day and in different ways. The most active beetles work in the heat of the day and make balls of dung, which they roll away, while the quieter tunnellers and the beetles that actually live in the dung heaps work at dusk.

MAKING THINGS WORK

Another way of avoiding competition in a habitat is for different groups of animals to be active in an area at different times of day. So in many habitats there are animals that feed and hunt during the day, while others, such as moths, bats, owls, and badgers, come out only at night.

Every community has a mixture of organisms to avoid too much direct competition. Although the competition that does take place can make life hard for animals and plants alike, it actually helps to prevent big changes that might result in far too many of one particular type of organism or species being wiped out in an area.

This American false vampire bat, like most other bats, only hunts at night to avoid competition with daylight hunters. It feeds on night-flying insects and its face and ears show many adaptations. These make the bat successful at flying, navigating, and feeding in the dark.

COMPETITION MATTERS!

Most of the time competition between different groups of animals and plants has a very positive effect. Many animals have very effective adaptations, which help them to compete more successfully. But without competition, things can go terribly wrong.

SCIENCE PIONEERS David Klein and the reindeer on St Matthew Island

During World War II, the US Coast Guard set up a navigation station operated by nineteen men on St Matthew, an unoccupied island 50 kilometres (31 miles) long and 6 kilometres (4 miles) wide in the Bering Sea off Alaska. The island was covered with a thick mat of **lichens**, and there were very few animals – just seabirds, nesting on the cliffs, a few species of ground nesting birds, one species of vole, and Arctic foxes.

In August 1944, the Coast Guard released 29 reindeer as a backup food source for the men, but the war ended soon after and the men left before they had the chance to shoot any reindeer. This left the reindeer with masses of food, no predators, and no competition. David Klein (then a biologist working for US Fish and Wildlife Service, now professor emeritus with the University of Alaska Fairbanks' Institute of Arctic Biology) first visited the island in 1957. He and his assistant Jim Whisenhant counted 1,350 very large, fit, healthy reindeer – quite an increase from 29! But already some areas of lichen were looking much the worse for wear, and lichen grows very slowly.

Klein did not return to the island again until summer 1963. As soon as he got to the shore all he could see were reindeer tracks, reindeer droppings, and lots of reindeer. The scientists counted 6,000 animals, and also observed the poor state of the lichens. There were simply too many reindeer. With no predators and no competition, the numbers just kept increasing.

Three years later, in 1966, Klein returned to St Matthew to find the island covered with skeletons. They counted only 42 live reindeer – 41 female adults, 1 male adult with abnormal antlers that probably could not breed, and no fawns. The reindeer population had dropped by 99 percent. By the 1980s, all the reindeer were dead and gone – the last few had died of old age – and the island returned to life as it had been for centuries. Without predators, the reindeer population had grown until it reached levels where the lichen was eaten far faster than it could be replaced. The reindeer literally ran out of food and eventually starved to death.

At the height of their population explosion there were 47 reindeer to every square mile of St Matthew. They were too successful for their own good, and without any competitors or predators to keep the numbers down, the reindeer died of starvation.

Protecting the environment

Scientists and governments have become more aware of the ways in which changes in the environment have an impact on the living things that make it their home. Anything that upsets the balance of organisms can change and perhaps damage the balance of a whole ecosystem. In many cases, it is the actions of people that cause the problems. When it comes to competition, or predation, people win every time.

THE PART PEOPLE PLAY

Without doubt, people cause problems for the environment. Burning **fossil fuels** releases **greenhouse gases** as well as acids and other **pollutants** into the atmosphere. **Pesticides** and **herbicides** are sprayed on crops to kill pests and weeds, but they also affect other animals and plants. People also cut down forests, dam rivers, and pour waste materials into the seas and oceans of the world.

But it is also very important to remember that in many countries the whole ecology is largely the result of human intervention. In the UK, for example, there is hardly any land at all which is not the result of human activity. Many of the woodlands and forests were planted centuries ago for kings to hunt in. The fields, the hedgerows, the railways, the canals, and the paths of many rivers have all been affected over the years by human activity. Many precious habitats have been created by people, and are now being changed by people again. Heathland habitats are becoming rare because fewer people keep sheep. Without sheep grazing, heathland soon stops being heathland because all the tree seedlings grow up and it starts to become woodland. The only way to keep that land as heathland is to put the sheep back.

CASE STUDY the collapse of Peruvian anchovy fishing industry

From 1960 to 1971, the Peruvian anchovy fishery was the biggest in the world, bringing in up to 12 million tonnes of fish every year. More and more people came to fish. But by 1972, the catch dropped to only 4 million tonnes, then less than 2 million, and by 1978 there was no catch at all. Overfishing caused the fishery to collapse. Overfishing means that too many fish are caught and there are not enough fish left in the sea to breed and replace the ones we eat. The Peruvian crash not only destroyed the fish stocks but also left 20,000 people without a way to earn a living.

People are very effective predators, but we do need to plan how much we take. Many different species of fish, from cod to anchovies, are threatened by overfishing. They simply cannot breed fast enough to keep up with our demands.

HOW TO PROTECT THE ENVIRONMENT

Around the world, people are starting to work together to protect the environment from damage. They are beginning to see the importance of **biodiversity**. If species of plants or animals become extinct, it has an effect on the whole habitat. Many animals and plants have adaptations that can be very useful to people. For example, several well-used medicines (such as aspirin, quinine, and tamoxifen, an anti-cancer drug) have been developed from chemicals found in plants. If plant species are lost before we have time to study them, medicines, which could save thousands of lives, could be lost as well. So protecting very rich environments with thousands of species, such as tropical rainforests and coral reefs, is very important.

As the world population grows and the planet becomes more crowded, we can see the need for **sustainable development**, where human progress is combined with a stable environment. As a result, complex food webs can be saved rather than destroyed.

When a habitat like a coral reef is damaged or destroyed, the animals and plants that live there cannot cope with the changes. They are adapted for life in one very specialized habitat and if it suddenly changes – the coral dies or is badly damaged by human activity – then they cannot adapt. Even if they survive, they are unlikely to breed successfully.

DOWN ON THE FARM

Farming can be very destructive, but many farmers are working hard to look after their soil and keep it fertile for years to come. Simple things like ploughing in the remains of crop plants and using animal waste as fertilizer as well as chemicals can help reduce the damage done by farming. Planting hedges, and leaving woodland standing not only stops soil erosion, it also helps many other species of animals and plants to survive.

RECENT DEVELOPMENT
the elk of Yellowstone Park

In the US there are a number of National Parks where the amazing countryside is managed so that it is as "natural" as possible. Unfortunately, in earlier times predators such as wolves were banned, so as a result the large herbivores like the elks have been able to breed very successfully. But the large numbers of elk, which now live in the park all year round, mean that the plant life is suffering. Aspen trees in particular are being badly affected because the elk love to eat the very young trees as they start to grow, so as old trees die there are no new ones to replace them. The challenge is to protect the elk and the trees at the same time. In 1995, a small number of wolves were reintroduced to Yellowstone Park in an attempt to restore the natural balance. Limiting the number of wolves makes sure that they do not kill too many elk.

Yellowstone may be a wilderness, but it is a carefully managed one. The magnificent Yellowstone park elk need a lot of effort to protect them and the other animals and plants that live in the same habitat.

ADAPTATIONS, COMPETITION – AND FEEDING THE WORLD!

Producing enough food for all the people on Earth is a big problem. Insects and pests eat lots of the crops while they are growing. Farmers often try to prevent pests by spraying chemicals to kill them, but these chemicals can cause problems for other members of the food webs. Scientists are trying to find biological ways of dealing with pests and problems rather than always using pesticides and herbicides. Sometimes they bring in new predators from other countries to kill the pests. For example, the Vedalia beetle was brought into California from Australia to control cottony cushion scale, a pest that was destroying citrus trees. It saved the citrus farms from destruction.

However, introducing a new species to an environment can be quite risky in case it competes too successfully with native organisms and drives them out. A better method is where local animals or plants are used to control a pest. This is called integrated pest management (IPM). Biological control and IPM have been very successful in many parts of the world.

Introducing a new animal has to be done with great care. Rabbits were introduced into Australia from the UK to provide a little sport. Before long they over-ran the country and were eating enough grass to feed millions of sheep. Then a disease, myxamatosis, was introduced to kill the rabbits. But the rabbits became immune to the disease and are still a massive problem.

SCIENCE PIONEERS the International Institute of Biological Control (IIBC)

The IIBC has had many success stories with integrated pest management. For example, rice is a vital food crop in Indonesia, but whole crops were being destroyed by brown planthoppers. Scientists from IIBC trained the rice farmers to identify the spiders that eat the planthoppers and conserve them. Using natural predators of the planthoppers, the pests were controlled, and the whole situation was turned around. In just three years, the money spent on pesticides had dropped by 90 percent, and yields of rice were increasing steadily.

WHEN THINGS GO WRONG

Sometimes an ecosystem suffers massive damage. It may be the result of a natural disaster, such as a hurricane or flood, or it could be caused by human error, for example oil spills from damaged tankers or poisonous waste released into rivers and seas. Many living organisms cope surprisingly well with natural disasters. There are always some things that are adapted for survival. But sometimes, particularly when people have caused the problems, it is not so easy. This is when it is very important to look for a solution based on the way animals and plants are adapted and make sure that the situation is made better, and not worse, by human interference.

Facing the future

As the 21st century progresses, new and different challenges are facing living things. Modern **biotechnology** makes it possible for scientists to choose adaptations for animals and plants, while the environment in which organisms live is changing faster than ever before.

GENETIC ENGINEERING – HIGH-SPEED ADAPTATION?

Genetic engineering, or genetic modification (GM), involves taking a gene from one organism and putting it into another, completely different one. Most adaptations introduced in this way make the organisms more useful to people. For example, bacteria have been engineered to make pure human insulin to help people with diabetes, while Tracy the **transgenic** sheep was given a human gene so she made a special protein in her milk, for people with rare blood clotting diseases who would otherwise bleed to death.

GM plants are much more common than engineered animals. Plants including cotton, potatoes, and tomatoes have been given a gene that makes them produce insecticide in their tissues. If a pest eats the plant, it is poisoned, so there

DNA is the molecule of life. It carries the design for every animal and plant on the planet. Now that people can change the DNA of animals and plants, the sky is the limit for the adaptations that can be produced.

RECENT DEVELOPMENTS Centre for Environmental Stress and Adaptation Research

Around the world, the environment is coming under threat from a range of stressful events, including land clearing, high salt levels, pesticides, industrial pollutants, and climate change from global warming. The Centre for Environmental Stress and Adaptation Research has been set up in Australia to find out how animals and plants respond and adapt in stressful environmental conditions. Research projects like this around the world will help us understand more about the living world and how it is coping with all the changes that are taking place.

is no longer any need for chemical spraying. Oil-producing plants, such as soya beans and oil seed rape, have been genetically modified to produce more oil, and other plants have been engineered so they do not ripen too quickly.

There are many advantages to GM plants and animals, but there may be problems too. One worry is that if GM organisms "escape" they may compete against natural species and win! Some scientists are working hard to develop better GM crops and to convince people that they are safe. Other scientists are convinced there are problems to come. Most of us just have to wait and see who is right. In some parts of the world, GM organisms are widely accepted, while in others, particularly in the UK and Europe, people have many concerns and the organisms are still very controversial.

One thing seems certain – the adaptations seen in the living world so far will be matched in future by the artificial adaptations produced by genetic engineering. How far they will spread into the natural world, and who will win the inevitable competitions between genetically modified and naturally adapted organisms, are questions that have yet to be answered. What is more, living organisms around the world need to adapt faster than ever to cope with the changes people are making. Who knows what the future will hold?

Further resources

MORE BOOKS TO READ

Barraclough, Sue, *Protecting Species and Habitats* (Franklin Watts, 2005)

Senker, Cath, *Charles Darwin* (Hodder Wayland, 2001)

Stockley, Corinne, *The Usborne Illustrated Dictionary of Biology* (Usborne Publishing, 2005)

Nature Encyclopedia (Dorling Kindersley, 1998)

USING THE INTERNET

Explore the Internet to find out more about adaptation and competition. You can use a search engine, such as www.yahooligans.com or www.google.com, and type in keywords such as *ecosystem*, *food web*, *natural selection*, *sustainable development*, or *transgenic*.

These search tips will help you find useful websites more quickly:

- Know exactly what you want to find out about first.
- Use only a few important keywords in a search, putting the most relevant words first.
- Be precise. Only use names of people, places, or things.

Glossary

adaptation special features of an organism that enable it to survive in a particular habitat

air pressure value of pressure of the air at any particular place

altitude height, typically measured as a distance above sea level

altitude sickness ill feeling caused when someone goes to a high altitude where there is less oxygen in the air

bacteria type of micro-organism that can be helpful, but that can also cause disease

biodiversity measure of the diversity of organisms living in a given area – both the different types of organisms and the variety within species

biosphere name for the whole world when it is considered as a single ecosystem

biotechnology science involving the use of biological processes and technological methods, carried out for our benefit

cellulose complex carbohydrate found in plant cell walls

chlorophyll chemical used by plants to capture the Sun's energy

chloroplast structure in the plant cell that contains chlorophyll

community all the populations of animals and plants that live together in a habitat at any one time

decomposer organism that breaks down natural waste, and dead plants and animals

DNA (deoxyribonucleic acid) molecule that carries the genetic code. It is found in the nucleus of the cell.

dormant in a deep sleep

ecosystem all the animals and plants living in an area, along with the interactions between the living organisms and the things that affect them such as the soil and the weather

embryo baby at a very early stage of development inside the mother

environment an organism's home and its surroundings

enzyme protein molecule that changes the rate of chemical reactions in living things without being affected itself in the process

evaporate turn from liquid into a vapour

faeces solid waste from the body

food chain links between different animals that feed on each other and on plants

food web model of a habitat showing how the animals and plants in different food chains are interconnected through their feeding habits

fossil fuels fuels formed over millions of years from the remains of ancient plants and animals. They are oil, coal, and natural gas.

fungi organisms that are neither plants nor animals – they do not move around and cannot photosynthesize

gene individual unit of information in the DNA

germinate begin to grow

greenhouse gas gas that adds to the greenhouse effect. This is the effect where the Earth's atmosphere heats up due to energy from the Sun being trapped.

habitat place where an animal or plant lives

herbicide chemical that kills plants (usually used on weeds)

hibernate spend time in a dormant state

innate in-built characteristic

insulate provide a barrier to temperature

invertebrate animal without a backbone

lichen simple plant that grows on rocks, walls, or trees

molecule group of atoms bonded together

mollusc group of invertebrates that have soft bodies and hard external shells

mutation change in the DNA

natural selection survival of the fittest organisms, and the passing on of their genes through reproduction

niche place and role taken by an organism within the ecosystem

nucleus central part of the cell, the nucleus contains the DNA

organism individual living thing, such as a plant or animal

osmosis special type of diffusion that involves the membrane, which is only permeable to certain molecules

ovipositor special organ for laying eggs

parasite organism which lives in or on another organism

pesticide chemical that kills pests, which are often insects

photosynthesis process by which green plants make food from carbon dioxide and water, using energy from the Sun

pollutant substance that pollutes the environment

predator animal that preys on other animals to obtain food

prey animal which is preyed on and eaten by predators

primary consumer first animal in the food chain. These are herbivores and omnivores.

producer first organism in the food chain

protist microscopic living organism, mainly single-celled

rehydrate to restore water

respiration using oxygen to release energy from food

ruminant animal that has a rumen, where food is taken first before going back to the mouth. This is known as chewing the cud.

secondary consumer second animal in the food chain. These animals eat the primary consumers.

species specific group of very closely related organisms whose members can breed successfully to produce fertile offspring

sustainable development development in harmony with the natural environment, which can be sustained without causing unnecessary damage to the environment

transgenic animal or plant that contains a gene from an organism of a different species, put in place by genetic engineering

tundra treeless Arctic region of Europe, Asia, and North America

urine liquid waste passed out from the body

vertebrate animal with a backbone

Index